How Many Do You See?

An English/Spanish Counting Book

Written by: Darlene Jack
Illustrated by: Justin Jack

ISBN: 978-1-7364150-4-7

Hi, my name is Lovely. Join me and my friends on our counting journey as we learn to count to 10. Our Earth has one Sun. It shines so bright. The Sun lights up the Earth with all of its might.

How many Suns do you see?

Count with me.

One /Uno

1

One
Uno

The next number on our journey is number two. I have two feet. Our feet help us walk around earth.

How many feet do you see?

Count with me.

One /Uno, two/dos,

2

Two
Dos

Three is the next number on our journey. I see three trees. Trees give us shade from the hot sun. They are great fun. You can use trees to play hide and seek with your friends and family. Sounds fun to me.

How many trees do you see?

Count with me.

One /Uno, two/dos, three/tres

3

Three
Tres

Near the sea, are those elephants I see?

The elephant is one of the

biggest animals on land.

Elephants are so grand.

How many elephants do you see?

Count with me.

One /Uno, two/dos, three/tres, four/cuatro

4

Four
Cuatro

Another grand animal is the mighty lion.

The lion is the king of the jungle.

There no denying when you hear his roar

on the jungle floor.

He also has a beautiful mane.

His mane fits his face like a picture frame.

How many lions do you see?

Count with me.

One /Uno, two/dos, three/tres, four/cuatro, five/cinco

5

Five
Cinco

The turtle is also an awesome creature.

Some turtles can live on the land and the sea.

Turtles have such versatility.

How many turtles do you see?

Count with me.

One /Uno, two/dos, three/tres, four/cuatro, five/ cinco, six/seis

6

Six
Seis

Another animal that can live on the land

and the sea is the crab.

Their claws make things easy to grab.

How many crabs do you see?

Count with me.

One /Uno, two/dos, three/tres, four/cuatro, five/

cinco, six/seis, seven/siete

7

Seven
Siete

Some animals can only live in the sea.

The sea horse is one of those animals

that need water to breathe.

How many sea horses do you see?

Count with me.

One /Uno, two/dos, three/tres, four/cuatro, five/

cinco, six/seis, seven/siete, eight/ocho

8

Eight
Ocho

The star fish is another animal that

only live in the sea.

They are such a beauty.

How many star fish do you see?

Count with me.

One /Uno, two/dos, three/tres, four/cuatro, five/cinco,

six/seis, seven/siete, eight/ocho, nine/nueve

9

Nine
Nueve

There are so many animals

that live in the sea.

Some of the creatures like to swim

together like a family.

Fish that swim together are

called a school of fish.

It's a sight you don't want to miss.

How many fish do you see?

Count with me.

One /Uno, two/dos, three/tres, four/cuatro, five/ cinco, six/seis, seven/siete, eight/ocho, nine/nueve, ten/diez

10

Ten
Diez

Counting brings so much joy to me.

It helps me answer how many?

How many elephants and how many trees?

Numbers are awesome and I have the

ability to count 1, 2, 3.

Count to 10 one more time with me.

One/uno, two/dos, three/tres, four/cuatro, five/cinco

Six/seis, seven/siete, eight/ocho, nine/nueve, ten/diez

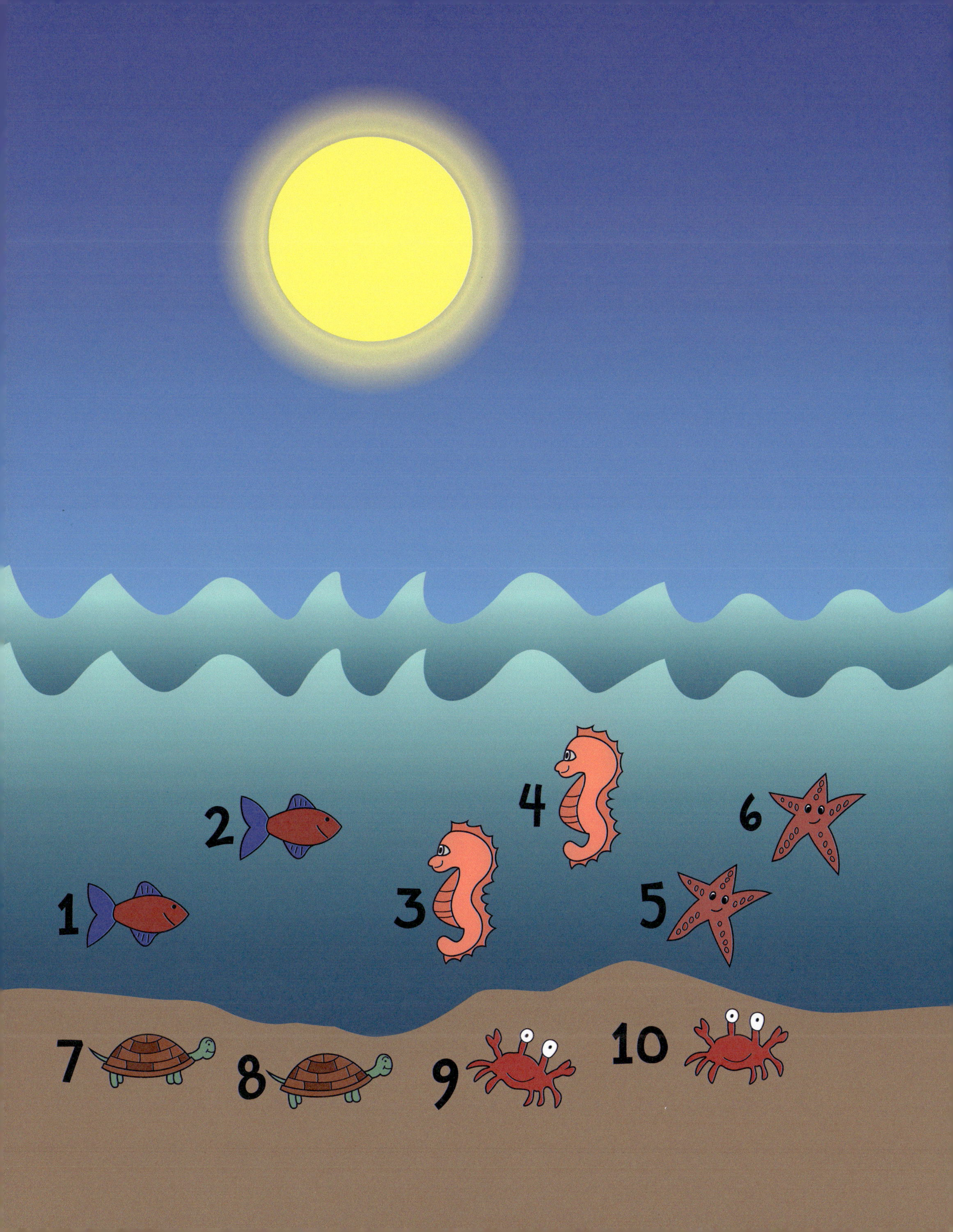
1
2
3
4
5
6
7
8
9
10

English	Spanish
One	Uno
Two	Dos
Three	Tres
Four	Cuatro
Five	Cinco
Six	Sies
Seven	Siete
Eight	Ocho
Nine	Nueve
Ten	Diez

Made in the USA
Monee, IL
05 November 2026

67386150R00019